ORYX SCIENCE BIBLIOGRAPHIES

Volume 5

Earthquake Prediction

by David A. Tyckoson

Oryx Press
1986

The rare Arabian Oryx is believed to have inspired the myth of the unicorn. This desert antelope became virtually extinct in the early 1960s. At that time several groups of international conservationists arranged to have 9 animals sent to the Phoenix Zoo to be the nucleus of a captive breeding herd. Today the Oryx population is over 400, and herds have been returned to reserves in Israel, Jordan, and Oman.

The Oryx Press
2214 North Central at Encanto
Phoenix, AZ 85004-1483

Published simultaneously in Canada

Printed and Bound in the United States of America

Library of Congress Cataloging-in-Publication Data

Tyckoson, David A.
Earthquake prediction.

(Oryx science bibliographies ; v. 5)
Includes index.
1. Earthquake prediction—Bibliography. I. Title.
II. Series.
Z6033.E1T9 1986 [QE538.8] 016.5512'2 86-2547
ISBN 0-89774-228-1

Table of Contents

About the Series

The *Oryx Science Bibliographies* are a new series of bibliographies designed to bring you the most recent references on the current issues in the sciences. Each issue will provide between 200-300 fully annotated references along with a "Research Review" covering the history and state of the art of the topic being covered. These bibliographies are intended to provide the student or researcher with an effective introduction to the hot topics in the sciences. Each bibliography also contains a number of special features, such as:

Evaluative Selection: The bibliographer reviews each article to ensure that only the most valuable references are included. The bibliography does not strive to be comprehensive but to include only the most important items on the topic.

Fully Annotated: All of the references are annotated, providing a useful summary of the source material for the user.

Readily Available Materials: All references chosen for inclusion are available at most libraries throughout the United States and Canada. Obscure sources that are difficult to obtain are usually avoided.

Highly Current: The bibliogaphy strives to include the most recent materials to keep it as up to date as possible. References from as recently as the last six months before publication may be found in all issues.

Key Articles Highlighted: The most important articles in each bibliography are highlighted in **boldface** so that the user interested in only a few materials can key in to those that are the most useful.

Research Review: Each issue contains a research review describing the state of the art of the subject being covered.

Undergraduate Level: Only materials that are written at the undergraduate student level are chosen. Neither highly technical nor extremely general articles are included.

English Language Only: Only English language materials are included.

Research Review: Earthquake Prediction

Parkfield, California: Sunday, January 17, 1988, 6:30 am. It is a cool winter morning in this desert town and most of the 38 inhabitants are sleeping quietly. The few that are awake hear a low rumbling sound from the north, almost as if a thunderstorm were approaching in the cloudless sky. Moments later, the ground begins to shake and, for the fourth time this century, Parkfield experiences a magnitude six earthquake. Because of its remote location, property damage is light and only one resident is injured. Following the earthquake, life in the town is back to normal except for the unusual behavior of farm animals and pets. Two days later, another quake strikes. The San Andreas fault is ripped apart, destroying Parkfield. The Parkfield earthquake has triggered the greatest quake to strike California since 1857. Two hundred miles to the south, the city of Los Angeles is devastated. Thousands are killed by the quake and subsequent fire and tens of thousands are injured. Three-fourths of the hospital beds in the city are unusable and most transportation, power, and water lines are severed. Despite recent changes in building codes, more than one-third of the buildings in the city sustain some damage. The greatest natural disaster in the history of the United States has occurred despite the efforts of scientists, government, and the public to prevent it.

Earthquake Prediction

Could this happen? Is southern California doomed to fall into the ocean? Can anything be done to prevent such a disaster? For the past twenty years, geologists have been attempting to answer these questions and to find some way to predict when an earthquake will occur. If earthquakes could be predicted in the same manner that we predict the weather, we could save lives and minimize property damage. Many different methods have been used to try to forecast earthquakes. Scientists examine the historical earthquake record for the region and also monitor such indicators as groundwater, gas emissions, gravity changes, slope, fault movements, electric currents, magnetic field strength, and animal behavior. However, none of these precursors is entirely accurate and most offer only a small indication of a forthcoming earthquake.

Definition of an Earthquake Prediction

For a forecast to qualify as an official prediction, three pieces of information must be provided: the location, the magnitude, and the time. If only the location and magnitude are given, it is termed an earthquake forecast. In order to be useful to the public, the prediction must contain an accurate designation of the time frame before the quake. However, accurately predicting the time of an earthquake has proven to be the most elusive factor. To achieve this goal, researchers are looking for accurate precursors to earthquakes.

Chemical and Hydrological Precursors

The most reliable indicators of earthquake activity have been changes in the level of the water table and in the chemical composition of groundwater. Because of the higher pressure within the earth prior to an earthquake, groundwater levels vary and more gases are squeezed out of the rocks. By measuring changes in the concentrations of various compounds, particularly radon, hydrogen, methane, and argon, scientists can predict earthquakes on a short term basis. However, not every quake exhibits the same behavior. Prior to most earthquakes, groundwater and radon levels increase, but in a few instances they have been observed to decrease. Although this method remains one of the most useful, further research is needed before earthquakes can be predicted on a regular basis.

Electromagnetic Precursors

Another method used to predict earthquakes is the use of electromagnetic fields and currents. Researchers have found that when rocks are subject to high pressure, a small electric current is generated. This electrotelluric current can then be monitored to determine if and when an earthquake will occur. It may also account for some related earthquake effects, such as earthquake lights and unusual animal behavior. Other scientists have theorized that these currents may be the source of electromagnetic radiation associated with earthquakes and that quakes could be predicted by monitoring radio waves.

Land Deformation and Tilt Measurements

Variations in the elevation and slope of the land around a fault are two of the long-term precursors to seismic activity. Researchers use both tiltmeters and laser ranging from spacecraft to monitor these changes. It has been shown that when a region experiences an uplift or severe change in slope, it is highly probable that an earthquake will occur. One of the most well-observed incidents of uplift occurred in the Mojave Desert at Palmdale, California. The Palmdale Bulge was seen to undergo an extreme uplift between the 1960s and 1980. Scientists are still debating whether the uplift is real or due to measurement errors in the survey. Whichever is the case, Palmdale is now one of the most heavily monitored earthquake areas in the world.

Stress and Seismicity

Two additional indicators of earthquake activity are the stress and seismicity of the earth along a given fault line. According to the theory of plate tectonics, as two plates move relative to one another they must slide along a fault line. Small earthquakes are generated whenever movement occurs. If the plates lock and no movement occurs, the stress inside of the earth builds until an earthquake relieves it. As the stress builds, the level of microearthquakes decreases. By monitoring the number and strength of small earthquakes, we can determine if a large earthquake will occur.

Seismic Gaps

One of the most accurate long-term earthquake predictors is the existence of a seismic gap, or a region where earthquake activity is far below normal. In an ideal seismic gap, the activity within the gap is at a very low level but remains normal in the surrounding region. This results in a doughnut pattern of seismic activity in which the center is a prime candidate for a future quake. This method has been used to successfully forecast earthquakes in Mexico and Alaska.

Animal Behavior

Reports of unusual animal behavior have been associated with earthquakes from antiquity to the present day. Usually these reports have been the results of anecdotal and unscientific observations and are reported only after the fact. The scientific community has recently begun to take such reports seriously and is trying to measure the existence and accuracy of animal predictions. Studies indicate that there are indeed some correlations between behavior and earthquakes, but not one that can be consistently used for prediction purposes. Scientists have theorized that animals may be able to detect earthquakes by hearing sound waves generated by rocks under pressure, sensing changes in the magnetic or electrical fields, or smelling noxious gases forced out of the earth. If we can determine how animals detect quakes, we could use those same methods to predict earthquakes and save human lives.

Social Aspects of Earthquake Prediction

Earthquake predictions have a tremendous effect on the economic, sociological, and political environment of an area. If an earthquake warning is issued, the people will decrease the amount of construction and spending, adversely affecting the economy. Some will permanently move from the region and others will not move in until after the earthquake occurs. Local government officials will try to minimize this damage and may not issue an earthquake warning even though a quake is imminent. If a warning is issued, the mental health of the region will decline and some unscrupulous individuals may take advantage of the situation for their own personal gains. Sometimes the damage to an area due to the prediction is worse than the damage caused by the quake itself. Business, government, and the public must do everything possible to maintain the economic and social health of a community in the face of a forthcoming earthquake.

International Earthquake Prediction Efforts

Four nations currently lead the world in

earthquake prediction: China, Japan, the Soviet Union, and the United States. Chinese scientists have had the greatest success at predicting earthquakes so far but have also suffered from some devastating failures. They saved thousands of lives in Haicheng, but hundreds of thousands perished in the Tangshan earthquake. Chinese scientists collect an enormous amount of data, often using volunteers, and then synthesize all of it to make a prediction. In Japan, Tokyo is long overdue for a major earthquake and an intensive seismic monitoring effort has been established. Soviet scientists have had some success by measuring seismicity and have established an extensive groundwater monitoring system. American scientists were the first to develop the seismic gap method and have adopted the techniques of the other nations.

Successes and Failures

Only two earthquakes in history have been predicted accurately in place, time, and magnitude: one at Haicheng, China and one in the Pamir Mountains of the Soviet Union. The Haicheng earthquake remains the most spectacular success, with Chinese scientists using data from each of the different precursors, from radon gas to animal behavior, to accurately predict the quake. One year later, another earthquake occurred in Tangshan that was not predicted, destroying the city and killing hundreds of thousands of people. Many other earthquakes around the world have been forecast, but to be of use to the public the prediction must include the time as well as the place.

One of the most publicized prediction failures was for three great earthquakes that were to have struck Peru in 1981. Brian Brady, a U.S. Geological Survey scientist, used a theory of earthquake prediction that used both traditional data and quantum mechanics to make a very specific prediction. The third of the set of three earthquakes was to have been the strongest in history. The U.S. National Earthquake Prediction Evaluation Council did not endorse the prediction, but it caused a great deal of concern in Peru. Many people left the country and it suffered heavily economically. Although the three earthquakes did not occur, we cannot be certain that they might not in the future.

The Future of Earthquake Prediction

Will the Parkfield scenario come true? Can anything be done to prevent such a disaster? For Parkfield and the rest of southern California, the question is not one of if, but one of when. A great quake is inevitable. The National Earthquake Prediction Evaluation Council has endorsed a prediction of a magnitude six earthquake for Parkfield within four years of January 1988 and the historical record indicates that a great earthquake could occur at any time.

Ideally, scientists will be able to accurately forecast these earthquakes and warn the public of any dangers. However, widespread accurate predictions are still at least ten years away. In the meantime, the public can be protected by erecting safer buildings and by being educated on emergency procedures. If scientists from around the world work together, perhaps the cities of Los Angeles, Tokyo, and Beijing can be spared the disaster that occurred in Mexico City. By sharing information, data, and procedures across national boundaries, geologists can work together to predict the most deadly of natural disasters.

Earthquake Prediction

1. Abelson, Philip H. "Tectonic Hazards." Science, v. 214, October 2, 1981, p. 11.

The Department of the Interior may eliminate the $16 million earthquake prediction program in the United States. Considering the possible damage a major earthquake could cause, this program must be continued.

2. Allen, Clarence R. "Earthquake Prediction." Geology, v. 11, December 1983, p. 682.

The optimism for earthquake prediction of five years ago has been tempered by recent problems. The failure of the v_p/v_s technique, the inability to predict two California earthquakes, and the recognition that most quakes are composite events have complicated prediction. The next ten years will see the successful short-term prediction of some quakes and the increasing accuracy of long-term earthquake hazard analysis.

3. "Astrologers, Dreamers, and Psychics: Can They Foretell?" Science Digest, v. 83, March 1978, p. 16.

Researchers are looking at earthquake predictions made by psychics to determine if there is any correlation between those predictions and actual earthquakes. If any relationship is found, the psychic predictions could aid scientists in identifying forthcoming earthquakes.

4. Atkinson, Barry K. "Earthquake Prediction." Physics in Technology, v. 12, March 1981, pp. 60-68.

Earthquake prediction has reached the point where limited, routine forecasting is just around the corner. Reliable prediction of earthquakes will require both substantial improvement in the conceptual modelling of the physics of the earth and improvements in the accuracy and range of monitoring instruments.

5. "Counseling on Quake Prediction." Science News, v. 117, March 1, 1980, p. 136.

The National Earthquake Prediction Evaluation Council recently met for the first time. The purpose of this group is to make recommendations to the Geological Survey concerning the probability and accuracy of individual earthquake forecasts.

6. Douglas, J. H. "Earthquake Research: Rethinking Prediction." Science News, v. 115, February 3, 1979, pp. 74-76.

Earthquake prediction is not as simple as it seemed five years ago. The dilatancy theory of rock behavior has been found not to apply to many earthquakes. While some predictions have been successful, many others have failed. Different types of earthquakes behave in different manners and researchers must determine which precursors are accurate and which are just background noise.

7. Golden, Frederick and Philip Faflick. "Predicting Quakes: A Shaky Start." Time, v. 116, December 8, 1980, p. 69.

Chinese scientists successfully predicted the Haicheng earthquake, but they also failed to predict the quake at Tangshan. Plate tectonics account for most earthquakes, but at locations in Italy and the Mediterranean it is still very difficult to predict because more than two plates are involved.

8. Gwynne, Peter. "Let's Make the Most of Our Faults." National Wildlife, v. 19, October/November 1981, pp. 34-39.

Chinese scientists successfully predicted an earthquake at Haicheng but failed to predict one at Tangshan. The methods used in earthquake prediction vary considerably and are not entirely accurate. A major earthquake will hit California soon and officials need to be ready for a potential disaster.

9. Huyghe, Patrick. "Earthquakes: The Solar Connection." Science Digest, v. 90, October 1982, pp. 72-75+.

One maverick researcher believes that earthquakes are triggered by solar flares. There seems to be a correlation between the occurrence of large solar flares and earthquakes throughout the world.

10. "International Views on Quake Prediction." Science News, v. 115, April 28, 1979, p. 279.

At a recent conference on earthquake prediction, the Chinese delegation presented data on the precursors to the great Tangshan quake of 1976. Japan has developed a useful radon monitor and the U.S. and the U.S.S.R. have adopted some of each other's methods. Each country makes earthquake warnings differently, depending on its government and culture.

11. Kalicki, Anne C. "Earthquake." SciQuest, v. 54, July/August 1981, pp. 12-17.

Scientists are looking for seismic gaps where future earthquakes might occur. However, predictions are often ignored or create other damage to the social welfare of the area. Two Geological Survey scientists have made very specific predictions for a set of earthquakes in Peru. Unless the cost of a disaster is greater than the cost of preparedness, most cities do not undertake preparedness programs.

12. Kerr, Richard A. "Earthquakes: Prediction Proving Elusive." Science, v. 200, April 28, 1978, pp. 419-421.

After initial successes, scientists have found that there is no simple answer to earthquake prediction. The main problem is the lack of a long, detailed monitoring record of earthquake behavior. Small earthquake swarms and radon gas anomalies are now considered the best precursors. Despite these problems, researchers believe that widespread, reliable predictions will someday be available.

13. Kerr, Richard A. "Prospect for Short-Term Earthquake Prediction." Science, v. 227, January 4, 1985, p. 42.

We are approaching the age when we can accurately make short-term earthquake predictions. The Parkfield section of the San Andreas fault will allow us to study and test various earthquake hypotheses.

14. Kerr, Richard A. "Prospects for Earthquake Prediction Wane." Science, v. 206, November 2, 1979, pp. 542-545.

A recent earthquake in northern California has caused doubts about earthquake prediction. Despite its location in a heavily monitored area, no obvious precursors were observed. We really know less than we thought we did about how earthquakes behave.

15. Kisslinger, Carl. "Practical Approaches to Earthquake Prediction and Warning." Eos, v. 65, June 12, 1984, p. 398.

A summary report on the sixth U.S.-Japan seminar on earthquake prediction.

16. Morgan, Michael. "Predicting the Quakes." Americas, v. 32, May 1980, pp. 9-11.

Researchers are trying to find ways to predict earthquakes and minimize the damage they cause. Japanese scientists have conducted some studies in Peru and have helped to revise building codes and reduce the loss of life. A Peru-Brazil-U.S. study is currently using lasers to measure movements in the crustal plates.

17. "Predicting Earthquakes: Imperfect Science." U.S. News and World Report, v. 89, December 8, 1980, p. 9.

An earthquake in southern Italy was not predicted despite efforts by Rome's National Research Council to increase prediction efforts. None of the usual indicators were seen before the quake.

18. Raleigh, C. Barry. "When the Earth Quakes." Technology Review, v. 83, August/September 1981, pp. 46-55.

By carefully monitoring pressure, Chinese scientists successfully predicted the Haicheng earthquake of 1975. Because the populace was well informed and trusted the prediction, they obeyed the government and thousands of lives were saved. In this country, we need to temper the urge to forecast with the knowledge of the damage done by an inaccurate forecast.

19. "Random Results from Earthquake Prediction." New Scientist, v. 79, July 27, 1978, p. 252.

The Geological Survey is monitoring earthquake predictions made by anyone outside of the field, from physicists to fortune tellers. The results of over 2500 predictions indicate that the results of these amateur forecasts are worse than would be obtained by random predictions.

20. "Recent Events Shake Quake Prediction Theories." Research and Development, v. 26, September 1984, pp. 83-85.

An earthquake that struck San Jose, California was not predicted by any earthquake experts. Scientists are good at finding places where earthquakes are likely to occur and are good at designing buildings to withstand earthquakes, but no one can accurately predict earthquakes on a short-term basis.

21. "Reliable Predictions Still Far in Future." *USA Today*, v. 113, June 1985, pp. 9-10.

Despite gaining a considerable amount of knowledge about earthquakes, scientists are no closer to being able to predict quakes than they were ten years ago. Neither geological nor animal precursors have proven reliable.

22. "Richter Scale--Here's How It's Calculated." *Science Digest*, v. 83, March 1978, p. 14.

An explanation of what the Richter scale means and how earthquakes are measured against it.

23. Rikitake, Tsuneji. "Classification of Earthquake Precursors." *Tectonophysics*, v. 54, 1979, pp. 293-309.

Much data have been collected on the accuracy and the lead time of various earthquake precursors. All of these precursors are analyzed to determine their predictive value.

24. Rikitake, Tsuneji. "Classification of Earthquake Prediction Information for Practical Use." *Tectonophysics*, v. 46, April 4, 1978, pp. 175-185.

Earthquake indicators can be classified into three categories: A. within several hours, B. within a few days, and C. within an unknown time period but not very far off. Various precursors currently in use are analyzed for their value in predicting earthquakes in each category.

25. Rikitake, Tsuneji. *Current Research in Earthquake Prediction*. Dordrecht, Netherlands: D. Reidel Publishing Co., 1981. 383p. ISBN 90-277-1133-X.

A review of earthquake prediction research, including geological and electromagnetic precursors and the sociological aspects of earthquake prediction.

26. Rikitake, Tsuneji. "Earthquake Forecasting and Warning." *Developments in Earth and Planetary Sciences*, v. 3, 1982, pp. 1-386.

Modern techniques for predicting earthquakes in Japan, China, the United States, and the Soviet Union are examined. Problems of communications and public reaction to earthquake warnings are reviewed along with relevant litigation.

27. Rikitake, Tsuneji. Earthquake Prediction. Amsterdam: Elsevier Scientific Publishing Co., 1976. 357p. (Developments in Solid Earth Geophysics, v. 9) ISBN 04-444-1373-1.

A fairly technical treatise on earthquake prediction through 1975. All types of methods are covered, including animal behavior, land deformation, seismic activity, magnetic and electrical effects, chemical precursors, and the socioeconomic effects of earthquake warnings.

28. Rikitake, Tsuneji. "Earthquake Prediction and Warning." Interdisciplinary Science Reviews, v. 3, March 1978, pp. 58-70.

A description of earthquake prediction programs in Japan, the United States, the Soviet Union, and China and a review of methods and consequences of earthquake prediction.

29. "Scientists Closing in on Ability to Pinpoint Event." Science Digest, v. 83, March 1978, pp. 12-17+.

Scientists believe that earthquake prediction is just around the corner. By studying plate tectonics and rock mechanics, precursors can be identified that will warn of a coming event. The most likely study areas are the seismic gaps which are areas where earthquakes are not observed as frequently as expected.

30. Shapely, Deborah. "Earthquakes: Los Angeles Prediction Suggests Faults in Federal Policy." Science, v. 192, May 7, 1976, pp. 535-537.

A prediction of a magnitude six earthquake for the Palmdale, California area has raised concern among everyone from geologists to politicians to evangelists. Monitoring of the area has been stepped up. A committee has been established to determine when official earthquake warnings should be announced. Often the damage from issuing an earthquake warning is greater than that of the quake itself.

31. Simpson, David W. "Earthquake Prediction." Nature, v. 286, July 30, 1980, pp. 445-446.

The optimism of the early 1970s has been tempered by recent complications in earthquake prediction. Premonitory effects have been found to vary widely with the tectonic environment and earthquake magnitude. The seismic gap method has been one of the most successful in the long term, but it cannot be used for mid-plate quakes. Water related effects have been the most reliable for short-term predictions.

32. Spall, Henry. "Earthquake Prediction Research of the Seismological Laboratory, California Institute of Technology." Earthquake Information Bulletin, v. 11, May/June 1979, pp. 95-101.

Some of the earthquake prediction studies in progress at the laboratory involve a study of microearthquakes along the San Andreas fault, a seismic analysis of the Oaxaca, Mexico, area, measurements of p-waves, and changes in seismicity, tilt, gravity, and radon emissions.

33. Suzuki, Ziro. "Earthquake Prediction." Annual Review of Earth and Planetary Sciences, v. 10, 1982, pp. 235-256.

A systematic review of earthquake prediction, including geologic, geodetic, seismologic, magnetic, electrical, and chemical precursors as well as the associated sociological problems.

34. Van Rose, Susanna. "The Next Quake Will Be..." Geographical Magazine, v. 52, October 1979, pp. 25-30.

Earthquake prediction has become a respectable science since Chinese seismologists have shown that warnings of forthcoming earthquakes can be measured and analyzed. However, no one has yet developed appropriate evasive actions except to develop earthquake-resistant buildings.

35. Ward, Peter L. "Earthquake Prediction." Reviews of Geophysics and Space Physics, v. 17, April 1979, pp. 343-353.

The United States has greatly increased its earthquake prediction program. Researchers are using land deformation, laser ranging, tiltmeters, magnetic field changes, electrical resistivity, radon gas emissions, and animal behavior to determine their values as earthquake precursors. Studies of the social and economic impacts of earthquake prediction are also under way.

36. Wellborn, Stanley N. "Is a Killer Earthquake Due to Hit the U.S.?" U.S. News and World Report, v. 95, November 14, 1983, pp. 74-76.

It is inevitable that a large earthquake will strike the United States. Such an earthquake could kill up to 23,000 people and cause $69 billion of property damage. An earthquake would also severely disrupt services such as transportation and health care and could also create an economic disaster.

37. Wesson, Robert L. and John R. Filson. "The U.S. Earthquake Prediction Program." Earthquake Information Bulletin, v. 13, September/October 1981, pp. 164-174.

An overview of the strategies and objectives of the earthquake prediction program in the United States.

38. Williams, Richard L. "Science Tries to Break New Ground in Predicting Great Earthquakes." Smithsonian, v. 14, July 1983, pp. 40-51.

The only two successful earthquake predictions in history have been in Haicheng, China and in the Pamir mountains of the Soviet Union. Scientists use changes in tilt, gravity, electrical conductivity, magnetic field strength, radon gas emissions, and laser ranging to assess quake potential. We can also do some long range planning through zoning and building codes to reduce possible damages.

Earthquake Research in China

39. Bennett, Gordon. "Mass Campaigns and Earthquakes: Haicheng 1975." China Quarterly, no. 77, March 1979, pp. 94-112.

The successful prediction of the Haicheng earthquake saved many thousands of lives. The use of amateur seismological study groups in helping to form the prediction is examined.

40. "The Chinese People and the Haicheng Earthquake." Technology Review, v. 83, August/September 1981, pp. 52-53.

The Chinese earthquake program is based on five basic principles: 1. A concern for the people affected, 2. A good track record in predicting earthquakes, which breeds confidence in the system, 3. The government is prepared to distribute relief items when necessary, 4. The government will help overcome the effects quickly to resume production levels in agriculture and industry, and 5. The present government will do much more than the pre-liberation government ever did.

41. "Coping with Quakes--The Chinese Connection." Earth Science, v. 36, Summer 1983, p. 10.

An overview of the Chinese earthquake prediction program.

42. Fang, Wang. "The 1976 Tangshan Earthquake." Earthquake Information Bulletin, v. 11, May/June 1979, pp. 106-109.

A description of the Tangshan earthquake and its precursors. Because the precursory phenomena differed from other Chinese earthquakes and there were no foreshocks, no prediction was made.

43. Gere, James M. and Haresh C. Shah. "A Visit to Tangshan, China." Earthquake Information Bulletin, v. 13, January/February 1981, pp. 4-11.

A report by the first American team to visit Tangshan, China since the great 1976 earthquake.

44. Hammond, Allen L. "Earthquakes: An Evacuation in China and a Warning in California." Science, v. 192, May 7, 1976, pp. 538-539.

The successful prediction of the Haicheng earthquake is described. The large amount of data collected helped to provide the basis for an accurate prediction. Although an earthquake warning has been issued for southern California, not enough data are being collected to provide a prediction matching the Chinese work.

45. Meredith, Dennis. "Earthquake Research and Political Tremors in China." Technology Review, v. 81, October 1978, pp. 15-18.

China provides an excellent opportunity for earthquake research, but politics may prevent this. Although there are a tremendous number of people monitoring earthquake precursors, they do not have the required theoretical background and may be collecting poor data. Chinese scientists also do not seem willing to share their research with the rest of the world.

46. Raleigh, C. B. et al. "Prediction of the Haicheng Earthquake." Eos, v. 58, May 1977, pp. 236-272.

The report of the U.S. study team investigating the prediction of the Haicheng earthquake. A description of the earthquake and the precursors used to determine its occurrence are provided as well as a description of the Chinese earthquake observation and warning system.

47. "Seismology: Automatic System." Beijing Review, v. 26, March 28, 1983, p. 30.

The first fully automatic seismic monitoring system in China is now operational. The system can record basic data about an earthquake and transmit it to a central location where it can be analyzed.

48. Shapeley, Deborah. "Chinese Earthquakes: The Maoist Approach to Seismology." Science, v. 193, August 20, 1976, pp. 656-657.

In China, seismology is an ideal people's science. Thousands of amateur teams record data and contribute to the earthquake prediction program. Chinese scientists monitor earthquake-prone areas in a similar manner to American scientists. When activity picks up, the public becomes involved through education and volunteer programs. Predictions can be made by a number of different governmental authorities.

49. Trotter, Robert J. "China Faces the Future: Seismology." Science News, v. 116, October 27, 1979, pp. 285-286.

China's earthquake program begins with geological studies and historical records. Active areas are put on a long-term prediction status and monitoring stations are established. If anomalies are found, an area can be upgraded to medium or short-term status, where it is intensely monitored.

50. Wallace, Robert E. "Tangshan, Six Years Later." Earthquake Information Bulletin, v. 15, May/June 1983, pp. 102-107.

A report of a visit to Tangshan as rebuilt following the 1976 earthquake.

51. Wallace, Robert E. and Tan-Liang Teng. "Prediction of the Sungpan-Pingwu Earthquakes, August 1976." Seismological Society of America Bulletin, v. 70, August 1980, pp. 1199-1223.

On August 16, 22, and 23, 1976, three large earthquakes were predicted in the Szechuan Province of China. The successful predictions saved thousands of lives. Over 1,300 observations by both scientists and lay brigades were used in predicting the events. Precursors included outgassing, fireballs and earthquake lights, abnormal animal and plant behavior, and telluric currents.

52. West, Susan. "Getting to the Source of Precursors." Science News, v. 119, January 10, 1981, p. 24.

The strength of the Chinese prediction program has been the large number of observations gathered and their analysis as possible precursors. The anomalies measured as precursors appear to be related to the stage of stress build-up.

Earthquake Research in Japan

53. Aki, Keiiti. "Progress in Japanese Earthquake Prediction Research." Earthquake Information Bulletin, v. 10, November/December 1978, pp. 199-203.

Earthquake prediction research began in Japan in 1962 and has grown steadily ever since. Japanese scientists are extensively monitoring the Tokai area, which is considered a high-risk region. The success of the program is due to a large amount of accurate data and a good understanding of the occurrence of earthquakes.

54. Bain, Iain. "Tokyo is Due for a Major Earthquake." New Scientist, v. 88, October 23, 1980, p. 230.

Tokyo has not been struck by a great earthquake since 1929. The next one is long overdue. The government is taking all necessary precautions to prevent a disastrous fire following a quake.

55. "Earthquake Prediction and Preparedness in Japan." Earthquake Information Bulletin, v. 15, July/August 1983, pp. 142-147.

Japan has integrated the scientific, engineering, and social aspects of earthquake prediction more thoroughly than any other nation. The methods used in Japan are recommended for use in the United States.

56. Lewis, John. "A New Schedule for Disaster." Far Eastern Economic Review, v. 110, October 3, 1980, p. 26.

Two magnitude six earthquakes hit Tokyo on September 24-25, 1980. Scientists are predicting a major earthquake sometime during the next 25 years. The government has established a network of sensors to detect an earthquake and is working on a system to warn and protect the public.

57. Okado, Y. "First Results from Japanese Network for Earthquake Prediction." Nature, v. 312, December 6, 1984, pp. 500-501.

Japan has built the most extensive earthquake monitoring network in the world. This network consists of 71 stations monitoring seismicity, tilt, groundwater, and chemical emissions.

Earthquake Research in the Soviet Union

58. Rich, Vera. "Early Warning System Works." Nature, v. 308, March 29, 1984, p. 395.

No one was killed in a major earthquake in the Soviet city of Gazli because the earthquake was predicted by the Soviet seismic warning network.

59. Rich, Vera. "Soviet Shake Up." Nature, v. 296, April 8, 1982, p. 485.

The Soviet Union has opened its first station in a national seismic early warning system. Primary indicators will be changes in the radon content of well water, changes in the magnetic field, and microdisplacement of crustal blocks. Animal behavior and local folklore will also be examined.

60. Sadovski, M. A. "Earthquakes Can Be Forecast." Science in the USSR, no. 5, 1981, pp. 58-71.

An overview of the earthquake prediction program of the Academy of Sciences of the USSR, including such techniques as measuring the piezoelectric effect and seismic zonation.

61. Simpson, David W. "Soviet Prediction of a Major Earthquake." Earthquake Information Bulletin, v. 11, November/December 1979, pp. 234-235.

Soviet scientists predicted an earthquake in the Tadjikistan-Kirghizia region in 1978. Intermediate precursors were geodetic and seismic anomalies and the short-term indicators included tilt, magnetic, and chemical anomalies.

62. Spall, Henry. "Earthquake Prediction in the Soviet Union--An Interview with I. L. Neresov." Earthquake Information Bulletin, v. 12, March/April 1980, pp. 60-63.

The Soviet prediction program began in 1950 and has grown considerably over the years. The first successes were with the v_p/v_s ratios. Later, seismic zones or gaps were identified as high probability areas. The current program is similar to the one in the United States.

63. Spall, Henry. "Earthquake Research in the Soviet Union." Earthquake Information Bulletin, v. 11, January/February 1979, pp. 22-25.

Soviet scientists have been working on earthquake prediction since 1949. They have developed the v_s/v_p method for predicting earthquakes, but have not had the same success that the Chinese have had.

64. "Successful Soviet Quake Prediction." Science News, v. 115, March 17, 1979, p. 169.

Using seismicity, tilt measurements, gravity changes, and well water levels, Soviet scientists have successfully predicted a magnitude 6.7 earthquake in the Alai mountain region.

65. Tinker, Jon. "Soviet Prediction of Iran Earthquake Comes True." New Scientist, v. 80, October 12, 1978, p. 91.

Soviet scientists predicted the Tabas, Iran earthquake 10-15 days before it occurred. Changes in groundwater levels and slope were the primary indicators. A canal project in the Soviet Kara Kum Desert has increased the chances of a major earthquake occurring along the Iran-USSR border.

Parkfield, California: The Next Southern California Earthquake

66. Adler, Jerry et al. "Forecasting Future Shock." Newsweek, v. 106, September 30, 1985, pp. 21-27.

The recent Mexico City earthquake again indicates what the damage to a highly populated area will be during a large earthquake. One such quake has been predicted for southern California during the next thirty years. Scientists have set up a network of sensors to monitor the situation. We can start protecting ourselves now by building earthquake-proof buildings and improving existing structures.

67. Anderson, Ian. "The Town Facing an Earthquake in 1988." New Scientist, v. 106, May 9, 1985, p. 9.

A magnitude 5.5-6.0 earthquake has been forecast for Parkfield, California within five years of January 1988. This is the first earthquake warning approved by the National Earthquake Prediction Evaluation Council. The Parkfield area is being extensively monitored for precursors to the event.

68. Begley, Sharon. "California's 'No Fault' Fault." Newsweek, v. 104, October 15, 1984, p. 111.

The San Andreas fault causes a major earthquake once every 145 years. There is a 50% to 90% chance of such a quake during the next fifty years.

69. "California's Fate." Time, v. 112, September 25, 1978, pp. 80-81.

A scientist at the California Institute of Technology has predicted a major earthquake along the San Andreas fault during the next fifty years.

70. David, Lester. "The Great Quake of 1991." Mechanix Illustrated, v. 77, October 1981, pp. 43-45+.

A fictional scenario of the devastation due to hit southern California if a major earthquake occurs. All of the damage reports are based on actual estimates by the Federal Emergency Management Agency.

71. DeLorenzo, Ron. "California Earthquakes: Predicting the Next Big One Using Radiocarbon Dating." Journal of Chemical Education, v. 57, August 1980, p. 601.

Radiocarbon data show that on the average, a major earthquake strikes California once every 160 years.

72. "Earthquakes: To Live and Shake in L.A." _Science News_, v. 128, November 30, 1985, p. 342.

A new report from the Geological Survey describes the hazard to the city of Los Angeles from earthquakes on the San Andreas and other faults.

73. "Earthquakes and Risk in California." _Earthquake Information Bulletin_, v. 14, May/June 1982, pp. 98-107.

By examining seismic gaps, researchers have identified the following probabilities for earthquakes along the San Andreas fault: 1. Low potential on the northernmost 270 miles, 2. Very low potential on the central 100 miles, 3. Significant hazard on the south central 220 miles, and 4. High risk on the southern 120 miles.

74. "Indicators Point to Big California Quake." _Industrial Research and Development_, v. 24, January 1982, pp. 59-60.

The last major earthquake to strike California was in 1857. By examining an excavated trench and using radiocarbon dating, earthquakes have been examined over the last eight centuries. Because the average interval between earthquakes is 145 years, an earthquake could occur at any time in the future.

75. Kerr, Richard A. "California's Shaking Next Time." _Science_, v. 215, January 22, 1982, pp. 385-387.

It has been 125 years since the last great California earthquake and scientists will be watching for precursors of the next one. The southern section of the San Andreas fault is the expected site, but it is not known how much damage will occur.

76. Kerr, Richard A. "Earthquake Forecast Endorsed." _Science_, v. 228, April 19, 1985, p. 311.

The National Earthquake Prediction Evaluation Council has endorsed the prediction that a magnitude 5.6 earthquake will occur near Parkfield, California in 1988.

77. Kerr, Richard A. "An Encouraging Long-Term Quake Forecast." Science, v. 225, July 20, 1984, pp. 300-301.

An analysis of the twists and turns in the Calaveras, California fault indicated that an earthquake might occur near Morgan Hills. On April 24, 1984, an earthquake did occur. By examining other places where fault lines bend, some earthquakes in northern California may be forecast.

78. Kerr, Richard A. "How to Catch an Earthquake." Science, v. 223, January 6, 1984, p. 38. Reprinted in: Earthquake Information Bulletin, v. 17, January/February 1985, pp. 33-34.

Geophysicists are confident that a magnitude 5.6 earthquake will occur at Parkfield around 1987-1988. This area is being extensively monitored to determine precursors and to test various earthquake prediction theories.

79. Kerr, Richard A. "Stalking the Next Parkfield Earthquake." Science, v. 223, January 6, 1984, pp. 36-38.

The Parkfield section of the San Andreas fault has been thoroughly studied and a magnitude 5.6 earthquake has been predicted for 1988. This part of the fault has been very well behaved and has followed a predictable pattern in the past.

80. Kerr, Richard A. "Stalking the Next Parkfield Earthquake in Central California." Earthquake Information Bulletin, v. 16, November/December 1984, pp. 215-224.

A nineteen-mile stretch of the San Andreas fault near Parkfield is one of the most well-monitored and well-understood faults in the world. Earthquakes of magnitude 5-6 occur regularly on this section. A prediction has been made for a quake of magnitude 5.6 to occur within 4.3 years of January 1988.

81. Lindh, Allan et al. "The Parkfield Prediction Experiment." Earthquake Information Bulletin, v. 11, November/December 1979, pp. 209-213.

Magnitude six earthquakes regularly strike Parkfield, California, with the last one occurring in 1966. This makes Parkfield a prime location for prediction studies. Slip and seismicity data both indicate a quake is likely. Because of these indicators, an extensive study has been set up to monitor the area in hopes of predicting the next Parkfield quake.

82. Mathewson, J. "Unearthing the Earth's History." Science News, v. 128, July 6, 1985, p. 5.

By examining the historical record of great earthquakes in California, researchers have found that the average interval between events is 150 years. Since the last large earthquake occurred in 1857, the area is due for another big one soon.

83. "Parkfield Looks Forward to Rock and Roll." Discover, v. 6, June 1985, pp. 11-12.

The National Earthquake Prediction Evaluation Council has forecast an earthquake for Parkfield, California sometime before 1993.

84. "Past and Future Shocks." Scientific American, v. 239, October 1978, pp. 92-102.

By studying the last great southern California earthquake in 1857, scientists have determined that if it occurred today, the damage to Los Angeles would not be very extensive. Tall buildings would be the most at risk due to the propagation of low frequency waves. Another great quake will occur sometime in the future.

85. Simon, C. "California's Quakes: Narrower Odds." Science News, v. 124, December 24, 1983, p. 404.

Scientists have predicted a 25% chance for a great earthquake along the southern third of the San Andreas fault in the next 20-25 years. This prediction is based on a statistical analysis of the time between earthquakes and the time elapsed since the last major quake.

86. Smith, R. Jeffrey. "Waiting for the Other Plate to Drop in California." Science, v. 201, September 1, 1978, p. 797.

Using laser satellite measurements, NASA scientists have predicted that a major earthquake will occur on the San Andreas fault sooner than expected. The measurements indicate that the plates are moving 50% faster than previously supposed. A recent magnitude five earthquake in Santa Barbara may be a precursor of the big one.

87. "Southern California's Last Great Earthquake and Clues About the Next One." Earthquake Information Bulletin, v. 14, July/August 1982, pp. 134-139.

The last great earthquake in California occurred in 1857. The average time between events is 125 years. The last quake was triggered by foreshocks and this may happen again if a sizeable earthquake occurs at Parkfield.

88. Weisburd, S. "Parkfield Quake Prediction Certified." Science News, v. 127, April 13, 1985, p. 228.

The National Earthquake Prediction Evaluation Council has endorsed an earthquake prediction for Parkfield within the next eight years. This is the first prediction authorized by the council.

89. Wesson, Robert L. and Robert E. Wallace. "Predicting the Next Great Earthquake in California." Scientific American, v. 252, February 1985, pp. 35-43.

No great earthquake has hit California since 1857. A description of the theory supporting the prediction of the Parkfield earthquake of 1988 is provided along with the possibility that it might lead to another great earthquake. The economic and social consequences of the next California earthquake are also discussed.

Earthquakes in Eastern North America

90. "Delaying the Inevitable." Macleans, v. 96, October 24, 1983, p. 48.

A recent earthquake that struck New York, Ontario, and Quebec increases the chances that a major earthquake will hit the St. Lawrence valley.

91. Dickinson, Terence. "Predicting the Eastern Canadian Quake." Macleans, v. 94, November 2, 1981, pp. 56-58.

There is a 40% chance of a magnitude seven earthquake in the St. Lawrence valley within the next ten years. Quakes in eastern Canada run along ancient fault lines, making exact predictions of locations and times impossible. An eastern quake can cause much more damage than a western quake because of the higher population density in the east.

92. Kerr, Richard A. "Assessing the Risk of Eastern U.S. Earthquakes." Science, v. 214, October 9, 1981, pp. 169-171.

Some of the most damaging earthquakes to strike the United States have occurred in the eastern and central region of the country and not along plate boundaries. Scientists have recently discovered the faults responsible for the 1811-1812 New Madrid, Missouri earthquakes and are conducting further research on earthquake hazards in the eastern U.S. Some new building codes have also been developed that should minimize damage from these earthquakes.

93. Kerr, Richard A. "U.S. Earthquake Hazards: Real But Uncertain in the East." Science, v. 201, September 15, 1978, pp. 1001-1003.

Major earthquakes have struck the eastern United States in the past and will strike again in the future. However, the possible damage from an eastern earthquake and the mechanisms behind those quakes are not well understood. Some studies indicate that quakes occur when old faults are reactivated, but the particular method remains unclear.

94. "Making Eastern Quake Plans." Science News, v. 125, April 28, 1984, p. 262.

A major earthquake of magnitude 7.5 or more may occur in the New Madrid, Missouri zone by the end of the century. The Federal Emergency Management Agency has recently begun planning for the aftermath of such an earthquake.

95. Milne, W. G. and M. J. Berry. "Trying to Reduce Earthquake Losses in Canada." Canadian Geographic, v. 100, April 1980, pp. 44-49.

Although most Canadian earthquakes occur in the west, some have also been reported along the St. Lawrence valley and in the far north. Scientists are attempting to learn how earthquakes operate in an attempt to predict future earthquakes.

96. "Quaking in the Midwest?" Newsweek, v. 103, April 30, 1984, p. 62.

Scientists have identified the fault associated with the 1811-1812 New Madrid, Missouri earthquakes and claim that another quake is overdue for the central United States.

97. U.S. House. Committee on Science and Technology. Earthquakes in the Eastern United States. Washington, D.C.: Government Printing Office, 1985. 774p. Superintendent of Documents Number Y4.Sci2:98/141.

The complete testimony before the House of Representatives on earthquake hazards and earthquake prediction in the eastern and central United States.

Electromagnetic Earthquake Precursors

98. Burton, Paul W. "Electrical Earthquake Prediction." Nature, v. 315, May 30, 1985, pp. 370-371.

Two Greek scientists have begun predicting earthquakes by using the electrotelluric currents in the earth. A network of sensors indicates that the current peaks before an earthquake and that an earthquake follows every peak.

99. Hoenig, Stuart A. "Aerosol Anomalies Preceding Earthquakes." Nature, v. 279, May 10, 1979, p. 169.

When rocks are compressed, a steady state current of up to 10^{-10} amperes is generated. As the rocks fail, both electrons and positive ions are emitted. This may support the charged ion theory of unusual animal behavior before earthquakes.

100. King, Chi-Yu. "Electromagnetic Emissions Before Earthquakes." Nature, v. 301, February 3, 1983, p. 377.

Anomalously high levels of electromagnetic emissions occurred thirty minutes prior to a magnitude seven earthquake in Japan. Two similar anomalies were identified in Japan and some have also been reported in the Soviet Union and China. This may be useful in forming short-term predictions.

101. Madden, Theodore R. "Electrical Measurements as Stress-Strain Monitors." Earthquake Information Bulletin, v. 11, January/February 1979, pp. 4-8.

Because the electrical properties of rocks are due to fluids in the pores and cracks, these properties should change when placed under pressure. Such variations should be very small, but may be useful precursors of earthquake activity.

102. "Radio Signals Before Quakes?" Science News, v. 121, March 20, 1982, p. 200.

An anomalous electromagnetic emission six days prior to a 1963 Chilean earthquake is hypothesized to have been due to stress release in quartz-bearing rocks. It may be possible to use broadcast-band radio receivers to aid in predicting earthquakes.

103. "Seismomagnetic Effects." Eos, v. 63, August 17, 1982, p. 609.

Recent research has demonstrated that magnetic effects related to earthquakes are much weaker than previously suspected. Detection of such changes will require very sensitive and favorably located instruments.

104. Stambolis, Costis. "Greek Physicists Perfect Earthquake Prediction." New Scientist, v. 98, May 12, 1983, p. 359.

Greek scientists have successfully predicted earthquakes over the last two years by monitoring the electrical pulses produced by the earth. When rocks containing an electrical charge are put under pressure, they emit an electric current. This method provides a seven to eleven hour warning of an impending earthquake.

105. "U.S.-China Quake Prediction Project." Science News, v. 118, December 6, 1980, p. 360.

The first joint U.S.-China earthquake prediction team will concentrate on gravity measurements as a predictive tool.

Chemical and Hydrological Earthquake Precursors

106. Brander, James. "Seismic Risk in Britain." New Scientist, v. 85, February 21, 1980, pp. 574-576.
Minor earthquakes occur throughout Great Britain every year. Most are not identified with any specific fault area. Groundwater circulation rates and radon content can be used to detect possible quake sites.

107. "Chlorine Clues to Quakes?" Science News, v. 128, August 17, 1985, p. 105.
Radon gas has frequently been used as a precursor of earthquakes, but it is difficult to monitor. Chlorine levels have also been seen to change before earthquakes, but they do not follow a standard pattern.

108. "Earthquakes, Volcanoes Give Chemical Signals." Chemical and Engineering News, v. 57, April 16, 1979, pp. 20-22.
Japanese scientists have determined that radon emissions are related to the amount of stress on rocks. This can be used to predict possible earthquakes and has been used successfully in both China and the Soviet Union.

109. Gold, Thomas and Steven Soter. "The Deep-Earth-Gas Hypothesis." Scientific American, v. 242, June 1980, pp. 154-165.
It is theorized that the hydrocarbons remaining from the formation of the earth are trapped in pockets inside the earth. As the gas slowly moves towards the surface, the added pressure causes earthquakes. If this theory is correct, methane emissions could be used to forecast earthquakes.

110. "Hydrogen May Flag Fault Movement." Science News, v. 118, November 1, 1980, p. 280.
Hydrogen gas may be released from the ground as movement occurs on an underground fault. Japanese scientists have discovered a wide variation of H_2 emissions along fault lines while observing no variation at sites distant from the fault. These emissions may provide a useful indicator of earthquake activity.

111. "Just Say When." Economist, v. 272, August 4, 1979, p. 69.

By using radon gas, Japanese scientists are monitoring indications of an earthquake in the Tokai region. Other gases may also be monitored to determine their predictive values.

112. King, Chi-Yu. "Do Radon Anomalies Predict Earthquakes?" Nature, v. 293, September 24, 1981, p. 262.

Although radon anomalies have been used by scientists in Japan, China, and the USSR for many years, American researchers have been skeptical. However, measurements in the Imperial Valley of California showed that radon production increased before the 1979 earthquake. While most radon anomalies produce an increase in gas flow, some show a decrease.

113. King, Chi-Yu. "Evaluating Hydrological and Geochemical Anomalies." Nature, v. 312, December 6, 1984, p. 501.

Scientists in the USSR are monitoring 240 wells at 83 sites for radon emissions. Forty parameters are measured, including water level, temperature, electrical conductivity, ion content, and gas pressure. It remains difficult to assess whether changes are due to a forthcoming earthquake or to other environmental factors.

114. King, Chi-Yu. "Geochemical Measurements Pertinent to Earthquake Prediction." Journal of Geophysical Research, v. 85, pt. B, June 10, 1980, pp. 3051-3121.

A collection of technical papers presented at the Fall 1978 meeting of the American Geophysical Union on the use of chemical techniques in predicting earthquakes.

115. King, Chi-Yu. "A Special Collection of Reports on Earthquake Prediction: Hydrologic and Geochemical Studies." Geophysical Research Letters, v. 8, May 1981, pp. 421-484.

A collection of fifteen technical papers presented at the Fall 1980 meeting of the American Geophysical Union.

116. "Natural Indicators." Beijing Review, v. 24, March 2, 1981, p. 28.

Underground water and gas emissions may become useful indicators for predicting earthquakes. Water level changes were observed before sixty earthquakes along with changes in the ratios of dissolved chemicals.

117. Oremland, Ronald S. "Methane in Association with Seismic Activity." Eos, v. 64, June 14, 1983, p. 410.

The upward flow of methane from the earth's core has been suggested as a possible cause of earthquakes. However, methane gas flow rates do not increase in proportion to the time of a forthcoming earthquake, contradicting this theory.

118. "Precision in Earthquake Prediction." Science News, v. 115, May 5, 1979, p. 297.

Describes a new radon monitor developed by the California Institute of Technology for use in earthquake prediction.

119. "Predicting Earthquakes." USA Today, v. 110, February 1982, p. 9.

Recent research has tied radon emission fluctuations to earthquakes. By measuring where an anomaly is observed, geologists can estimate the magnitude and location of the coming quake.

120. "Radon: A Message in Emissions?" Science News, v. 124, September 24, 1983, p. 206.

Radon measurements need to be correlated with other indicators in predicting earthquakes. In one case, a radon increase was followed by a major earthquake. In a second case, radon increases were correlated only with changes in groundwater levels and no earthquake occurred.

121. "Radon: Cause for Interest, Not Alarm." Science News, v. 120, November 21, 1981, p. 333.

Although much publicity has been generated due to increased radon levels in southern California, scientists are not predicting a major earthquake. Intense swarms of minor earthquakes have been observed and researchers are also beginning to monitor levels of hydrogen, helium, carbon dioxide, and methane.

122. "Radon: Clue to Earthquake Magnitude." Eos, v. 62, January 6, 1981, p. 2.

Changes in the flow of radon can be used to predict the magnitude of a forthcoming earthquake. The greater the distance from the epicenter that the changes are observed, the stronger the earthquake.

123. "Radon in California: Watching the Detectors." Science News, v. 120, October 17, 1981, p. 247.

An increase in radon content in southern California soil may be an indication that an earthquake is going to occur. Groundwater levels have also been rising recently.

124. "Radon Leaks Help to Predict Earthquakes." New Scientist, v. 100, October 27, 1983, p. 265.

Increases in radon emissions have preceded earthquakes in China, the Soviet Union, Japan, and California. Radon emissions increase when small cracks open in the earth prior to a quake. When used along with other methods, radon can be an accurate indicator of earthquake probability.

125. "Sensor Network to Predict Earthquakes." New Scientist, v. 105, March 21, 1985, p. 17.

A few days before a seismic event, hydrogen emissions increase. A network of fifteen sensors is being used to monitor hydrogen emissions. The theory behind the hydrogen is that plastic or serpentile rocks are forced into small cracks. The serpentile rock is formed by water reacting with rocks rich in iron and magnesium, with hydrogen produced as a by-product of the reaction.

126. "Shaking the Radon Out of the Earth." Analytical Chemistry, v. 51, October 1979, pp. 1183A-1186A.

Radon measurements are made by comparing the amounts of radon-222 and radon-220. Of eleven earthquakes of magnitude 2.0 or higher within 25 km of the recorder, three were preceded by radon anomalies and four showed possible radon variations.

127. Sugisaki, Ryuichi. "Changing He/Ar and N_2/Ar Ratios of Fault Air May Be Earthquake Precursors." Nature, v. 275, September 21, 1978, pp. 209-211.

Rather than rely solely on the amount of argon released in a seismic area as an earthquake predictor, the ratios of helium and nitrogen to argon should be used. In an experiment in Japan, these ratios peaked immediately before a quake.

128. Sugisaki, Ryuichi. "Deep-Seated Gas Emissions Induced by the Earth Tide." Science, v. 212, June 12, 1981, pp. 1264-1266.

The helium/argon ratio was measured in gas bubbles in water from a mineral spring along a fault line. Deep-seated gases characterized by higher helium/argon ratios are squeezed out by stress before an earthquake.

129. Thompson, Roland B. "Afloat on a Sea of Methane?" Technology Review, v. 81, October 1978, pp. 14-15.

One theory that would account for brightening skies, noxious odors, and unusual animal behavior before earthquakes is that there is a large amount of methane gas distributed throughout the earth's core. When cracks occur before an earthquake, some of this gas escapes.

130. Wakita, Hiroshi et al. "Radon Anomaly: A Possible Precursor of the 1978 Izu-Oshima-Kinkai Earthquake." Science, v. 207, February 22, 1980, pp. 882-883.

Changes in radon concentrations were observed in groundwater prior to a Japanese earthquake. The levels dropped suddenly before the quake and rose significantly afterwards.

Earthquake Prediction by Land Movement and Deformation

131. Alexander, George. "Quake Watch." Science 82, v. 3, September 1982, pp. 38-45.

Based on observations in China, Japan, and California, a theory of earthquakes based on deformation fronts is presented. In this theory, a stress wave passes through an area at a rate of 25-70 miles/year. As the wave passes through, precursors such as radon anomalies and land deformation are observed and small tremors are suppressed. After the wave passes, changes in strain along fault lines could trigger earthquakes.

132. Hindley, Keith. "Earthquake Waiting to Happen?" New Scientist, v. 77, January 26, 1978, pp. 228-230.

The recent elevation of an area along the San Andreas fault near Palmdale, California has alerted scientists to the possibility of an earthquake in southern California. A survey team will be constantly monitoring the bulge for activity.

133. Hunt, John. "Massive Releveling Project May Aid in Earthquake Predictions." Public Works, v. 110, August 1979, p. 76.

A massive leveling survey was conducted of the Palmdale Bulge to determine the exact extent of altitude changes. By coordinating these data with changes in local gravity strength, scientists hope to be able to determine if and when an earthquake will occur.

134. Kerr, Richard A. "Palmdale Bulge Doubts Now Taken Seriously." Science, v. 214, December 18, 1981, pp. 1331-1333.

Many scientists are beginning to have doubts about the existence or extent of the Palmdale Bulge. They believe that errors in measurement may account for most of the observed change in elevation. Although most agree that some uplift did occur, the corrected smaller changes are much easier to explain using current theories.

135. "NASA Probes What Makes Earth Quake." Design News, v. 40, April 9, 1984, pp. 33-34.

NASA is using laser ranging by satellite and very long baseline interferometry of quasars to measure movements along fault line and predict possible earthquakes.

136. "Palmdale Bulge Still a Puzzle." Eos, v. 61, July 1, 1980, pp. 515-516.

Researchers are still debating whether the Palmdale Bulge exists and what its meaning might be. Measurement errors may account for most of the observed uplift.

137. "Quakes and Quasars." Science Digest, v. 89, June 1981, p. 94.

Scientists are using quasars, which are so distant that they will remain fixed points of reference for centuries to come, as reference points for measuring the movement of the earth along fault lines.

138. Rundle, John B. and Marcie McNutt. "Southern California Uplift--Is It or Isn't It?" Eos, v. 62, March 10, 1981, pp. 97-98.

There is no clear decision yet on the existence of the Palmdale Bulge. Some scientists believe that it is an actual phenomenon and others believe that it is a result of survey errors.

139. "Space Data May Aid Earthquake Forecasts." Aviation Week and Space Technology, v. 109, July 10, 1978, p. 48.

Spacecraft laser ranging experiments are being used to monitor movement along the San Andreas fault.

140. Spall, Henry. "Space Techniques for Measuring Crustal Deformation." Earthquake Information Bulletin, v. 11, January/February 1979, pp. 9-17.

Laser ranging and very long baseline interferometry can both be used to detect movements along fault lines. Although it will take some time to determine plate motions and deformation, these data may be useful for earthquake prediction.

141. "Teams Map Bulge in Search of Quake Answers." Engineering News-Record, v. 200, January 19, 1978, p. 41.

Survey teams are measuring the Palmdale Bulge in the Mojave Desert to determine the exact extent of the uplift. The Geological Survey believes that the uplift may be an indicator of a coming quake.

142. "Telemetry to Predict Earthquakes." New Scientist, v. 105, March 21, 1985, p. 53.

The Japan National Research Center for Disaster Prevention uses a telemetry network to monitor earth movements along fault lines.

Monitoring the Stress and Seismicity of the Earth

143. Borok, V. I., L. Knopoff, and I. M. Rotvain. "Bursts of Aftershocks, Long-Term Precursors of Strong Earthquakes." Nature, v. 283, January 17, 1980, pp. 259-263.

Earthquakes that are followed within a short time by abnormally large numbers of aftershocks are hypothesized to be long-term precursors of stronger earthquakes. Eighteen of twenty-three strong earthquakes worldwide were predicted retrospectively using this criteria.

144. "CAD Assists in Earthquake Prediction." Civil Engineering, v. 54, February 1984, pp. 18-20.

Based on studies of Mt. St. Helens, a company has designed some new computer software that will analyze and plot seismic data. By plotting past seismic patterns in three dimensions, future quakes may be predicted.

145. Clark, Gregory A. and Peter W. Rodgers. "Adaptive Prediction Applied to Seismic Event Detection." Proceedings of the IEEE, v. 69, September 1981, pp. 1166-1168.

The Widrow-Hoff LMS adaptive filter can be used to provide useful data when detecting small seismic events from the micro-seismic background noise.

146. Crampin, Stuart et al. "Observations of Dilatancy-Induced Polarization Anomalies and Earthquake Prediction." Nature, v. 286, August 28, 1980, pp. 874-877.

Much of the search for earthquake precursors has relied on the fact that rocks under high stress will show dilatancy, the formation of stress-induced cracks. Seismic waves propagating through cracked material have a tendency to polarize along the alignment of the cracks. This may be used to monitor the state of dilatancy in a seismic region.

147. "Earthquake Monitoring Made to Measure." New Scientist, v. 82, May 17, 1979, p. 549.

New developments in technology have created a seismometer the size of a beer can. It is as sensitive as the larger models but not nearly as expensive.

148. "Earthquake Predictions Questioned." Science Digest, v. 90, August 1982, p. 20.

Earthquake predictions based on a reduction in small earthquake frequency may be in error. Since 1968, the Department of Defense has been closing down its monitoring stations, thus reducing the amount of data available and creating an artificially low seismic record.

149. "Earthquakes: The Calm Before the Storm." New Scientist, v. 79, July 13, 1978, p. 110.

All of the major earthquakes occurring in the Assam region of India over the past 150 years have been preceded by extensive periods of no seismic activity. The length of the quiet period was directly related to the magnitude of the quake. The current quiet period may be an indicator of an upcoming major earthquake.

150. Eaton, Jerry P. "Changes in the Seismicity of Central California." Earthquake Information Bulletin, v. 11, November/December 1979, pp. 205-208.

The epicenters of microearthquakes in central California are closely related to the principal faults in the area. They also suggest that the boundary between the Pacific and North American plates is much wider than expected. Further recording of microearthquakes will allow scientists to use these data to predict larger quakes.

151. "Experts Shaken." Economist, v. 274, March 1, 1980, p. 85.

A recent series of small earthquakes in California has provided researchers with no new clues to predicting the next great San Andreas earthquake.

152. Haberman, R. E. and Max Wyss. "Earthquake Triggering During Preparation for Great Earthquakes." Geophysical Research Letters, v. 11, April 1984, pp. 291-294.

Seismic quiescence has been proposed as a precursor to many large earthquakes. Two instances of smaller earthquakes triggering a great earthquake are examined. For this to happen, the stress in the area must be close to failure and the triggering quake must be large enough to affect the stress field but not large enough to cause an immediate main shock.

153. Kerr, Richard A. "Concern Rising About the Next Big Quake." Science, v. 207, February 15, 1980, pp. 748-749.

Instruments measuring the strain on the San Andreas fault have shown that the fault is now being pulled apart rather than squeezed together. It is uncertain what this implies for a future earthquake. Scientists will be measuring the strain as well as other possible precursors, such as radon emissions and groundwater levels.

154. Lee, W. H. K. and S. W. Stewart. "Microearthquake Networks and Earthquake Prediction." Earthquake Information Bulletin, v. 11, November/December 1979, pp. 192-195.

Microearthquakes of magnitude three or less have been used to study many seismological problems, including the stress levels preceding a stronger earthquake. At present we are collecting data on long-term microearthquake levels so that variations may be mapped against those of the future.

155. McEvilly, T. V. and R. Clymer. "Monitoring Seismic Wave Velocities In Situ." Earthquake Information Bulletin, v. 11, November/December 1979, pp. 214-220.

Reports from the Soviet Union indicate that the v_p/v_s ratio may be useful in predicting earthquakes. This method was tested using mobile recording stations and waves generated by explosions at a quarry. The results are inconclusive, but do show some promise for future studies.

156. Page, Robert A. "Estimating Earthquake Potential." Earthquake Information Bulletin, January/February 1980, pp. 16-24.

Both historic seismicity and the geological record are useful in predicting earthquake magnitudes. Estimates of earthquake potential are more reliable in regions of high seismicity than in relatively quiet regions.

157. Raleigh, C. B. et al. "Forecasting Southern California Earthquakes." Science, v. 217, September 17, 1982, pp. 1097-1104.

Earthquake frequency has increased in southern California over the last four years. The increased seismicity is associated with a marked deviation in the pattern of strain accumulation. The San Andreas fault has accumulated sufficient displacement to generate a great earthquake along part or all of its length.

158. Sherburne, Roger W. "Earthquake Predictions Using Seismic Velocity Ratios." Earthquake Information Bulletin, v. 11, January/February 1979, pp. 18-21.

Anomalies in the ratio of the velocities of compressional waves (p-waves) and shear waves (s-waves) may be used to predict earthquakes. This method has been successfully used in the Soviet Union, but has not had much success in California. It should be studied for its possible use in relation to other precursors.

159. Smith, Peter J. "Reservoirs and the Triggering of Earthquakes." Nature, v. 295, January 7, 1982, p. 9.

Dams and reservoirs can induce seismicity in areas where it was previously lacking. The weight of the water compresses the underlying rocks. As it seeps into the ground, it can change the coefficient of friction and also change the chemical and physical properties of the soil. As an example from the Soviet Union indicates, although reservoirs can cause earthquakes, the resulting pattern of seismicity depends on local conditions.

160. Smith, Warwick D. "The B-Value as an Earthquake Precursor." Nature, v. 289, January 15, 1981, pp. 136-139.

The b-paramater measures the relative numbers of small and large earthquakes. A systematic study of b-values in New Zealand has shown that within the vicinity of large forthcoming earthquakes there is initially an increase in b-value and then a return to normal. Earthquakes in Caracas and San Fernando also exhibit this pattern.

161. "Studies May Explain More About Earthquakes." Earth Science, v. 35, Spring 1982, pp. 35-36.

Researchers are studying the deep crust and upper mantle of the earth to detect possible earthquake causes. They are currently mapping the deep crust using seismic-reflection profiling.

162. Thatcher, W. "Seismic Triggering and Earthquake Prediction." Nature, v. 299, September 2, 1982, pp. 12-13.

Although most models assume that seismic stress is built up uniformly, this is not always so. Once such an area has been identified, an increase in microseismicity would indicate an increase in seismic risk. Another indicator is the magnitude of the stress or strain across the fault.

163. "Torturing Rocks to Protect People." USA Today, v. 110, February 1982, p. 9.

By putting rocks under extreme pressure and seeing how they fail, researchers are trying to find some indicators of earthquake behavior. They are also injecting water into the rock and examining any chemical changes that occur.

164. "Tsunami Warnings from Tuning in to Deeper Vibes." New Scientist, v. 78, June 22, 1978, p. 825.

Earthquakes that produce tsunamis, or large tidal waves, are characterized by an overwhelming amount of the energy being released as long-wavelength seismic waves. Three seismic stations in Hawaii, the Soviet Union, and Alaska are now monitoring these wavelengths in hopes of predicting tsunami producing earthquakes.

165. "A Very Predictable Kind of Earthquake." New Scientist, v. 74, April 7, 1977, p. 18.

A small earthquake was successfully forecast along the Calaveras, California fault during 1976. This may imply that other similar strike-slip faults can be predicted on a regular basis by monitoring seismic activity.

166. West, Susan. "Earthquake Studies: In the Lab and in the Field." Science News, v. 116, November 17, 1979, p. 344.

Researchers in the laboratory have discovered that the characteristics of foreshocks change before a laboratory-induced earthquake. This may be very useful for predicting natural earthquakes. This pattern was observed in China before the 1975 Haicheng earthquake.

167. Wyss, Max and M. Baer. "Seismic Quiescence in the Western Hellenic Arc May Foreshadow Large Earthquakes." Nature, v. 289, February 26, 1981, pp. 785-787.

The seismicity rate south of Crete was constant between 1950 and 1978, but in the western Hellenic trench the rate decreased by 80% after 1962. This may indicate that a major quake is due in that area.

168. Zoback, Mark D. and John C. Roller. "Magnitude of Shear Stress on the San Andreas Fault: Implications of a Stress Measurement Profile at Shallow Depth." Science, v. 206, October 26, 1979, pp. 445-447.

A profile measurement of shear stress perpendicular to the San Andreas fault near Palmdale, California shows a marked increase in stress with distance from the fault. By monitoring the stress build-up, it may be possible to predict future earthquakes in this area.

169. Zoback, Mark, John Roller, and John Healy. "In-Situ Measurement of the Earth's Stress Field." Earthquake Information Bulletin, v. 10, November/December 1978, pp. 214-219.

By examining changes in stress within the earth's crust, earthquakes might be predicted. A method for measuring this stress by drilling boreholes is described.

Seismic Gaps

170. Alexander, George. "Forecasting Earthquakes." Popular Science, v. 215, September 1979, pp. 79-81.

Analyzing seismic gaps has resulted in the successful forecast of earthquakes in the Soviet Union, Alaska, Japan, and Mexico. When an area is determined to be overdue for an earthquake and the surrounding regions are behaving normally, the anomaly is designated as a seismic gap. The most successful prediction using this method so far has been in the Oaxaca gap, but scientists are monitoring at least 25 other gaps around the world.

171. "Calm Before." Economist, v. 270, March 31, 1979, p. 72.

By investigating seismic gaps, scientists are more likely to forecast an earthquake than by using other methods. Two such forecasts have recently been made: one for Mexico and one for Alaska.

172. "Forecast of Mexican Quake Accurate, But Ignored." Science News, v. 114, December 9, 1978, p. 404.

Two University of Texas scientists predicted the location of a Mexican earthquake to within a mile and were very close to predicting the magnitude. They used the seismic gap method, in which a marked decrease in seismic activity precedes a major earthquake. Unfortunately, no one listened to the warning and there were no measurements taken to determine the time of the quake.

173. "Forecasting of Earthquakes Shows Gains." U.S. News and World Report, v. 86, April 30, 1979, p. 65.

Despite the devastation of recent earthquakes, scientists are beginning to be able to predict when and where they will strike. Earthquakes have been forecast successfully in Mexico and China. Efforts are currently under way to identify potential zones where a quake might occur.

174. "Growing Confidence in Earthquake Prediction." New Scientist, v. 82, June 7, 1979, p. 798.

Scientists from Columbia University have predicted that an earthquake will occur along the eastern gulf of Alaska 230 miles east of Anchorage. The date of the forthcoming quake is not known.

175. Harlow, David H. et al. "Quiet Zone Within a Seismic Gap Near Western Nicaragua: Possible Location of a Future Large Earthquake." Science, v. 213, August 7, 1981, pp. 648-651.

A seismic gap has been located off the coast of Nicaragua. This region is predicted to produce an earthquake of magnitude larger than seven, but the time frame is unknown.

176. Kerr, Richard A. "Another Successful Quake Forecast." Science, v. 203, March 16, 1979, p. 1091.

A recent Alaska earthquake occurred in the gap that scientists had forecast was due for a quake. However, the earthquake released the tension only in the eastern half of the gap, creating additional strain and the possibility of an even greater earthquake.

177. Kerr, Richard A. "Earthquake Prediction: Mexican Quake Shows One Way to Look for the Big Ones." Science, v. 203, March 2, 1979, pp. 860-862.

The recent Oaxaca, Mexico earthquake was predicted by using the seismic gap method. Seismic gaps are areas in which a quake is overdue and which are surrounded by areas of normal seismic activity. These regions are the most likely sites for future earthquake prediction studies.

178. Kerr, Richard A. "Harbingers of the Coalinga Earthquake." Science, v. 222, November 25, 1983, p. 918.

In retrospect, the Coalinga, California earthquake could have been predicted by the doughnut pattern of previous earthquakes around the area. When this pattern is observed, it implies that the center is under a great deal of stress. Another doughnut pattern has been identified, but scientists are unsure of its importance.

179. Lahr, John C. and George Plafker. "Holocene Pacific-North American Plate Interaction in Southern Alaska: Implications for the Yakataga Seismic Gap." Geology, v. 8, October 1980, pp. 483-486.

An earthquake at St. Elias, Alaska on February 28, 1979 ruptured only a fraction of the seismic gap in the region. A model of neotectonic deformation predicts a future magnitude eight earthquake between Icy Bay and the Aleutian Trench.

180. "Make a Date for the Next Alaska Shake." New Scientist, v. 91, July 16, 1981, p. 148.

Columbia University scientists have predicted a magnitude 8.0 earthquake for Cape Yakataga, Alaska. Ground damage could occur over 78,000 km^2 and dangerous tsunamis may be generated. The Geological Survey has installed seismographs and tilt-meters to monitor the area.

181. McCann, William R., Omar J. Perez, and Lynn R. Sykes. "Yakataga Gap, Alaska: Seismic History and Earthquake Potential." Science, v. 207, March 21, 1980, pp. 1309-1314.

A 250 km gap around Cape Yakataga in Alaska has shown little seismic activity, but has been surrounded by normal activity. A shock of magnitude 7.7 in 1979 ruptured only a small part of the gap, leaving the possibility of a major earthquake in the remainder.

182. McNally, Karen C. "Seismic Gaps in Space and Time." Annual Review of Earth and Planetary Science, v. 11, 1983, pp. 359-369.

The seismic gap concept has proven very successful in forecasting the sizes and locations of large, shallow earthquakes. Seismic gaps are defined, their relationship to plate tectonics is reviewed, the breaking of gaps is discussed, and suggestions for future research are provided.

183. "Oaxaca Quake Was Trapped After All." Science News, v. 114, December 16, 1978, p. 422.

Very small portable seismographs were used to measure seismic activity before and during the Oaxaca, Mexico earthquake. This quake had been predicted by the seismic gap method.

184. "Quakes to Come." Science Digest, v. 90, January 1982, p. 18.

A major earthquake has been predicted for Cape Yakataga, Alaska within the next 20-30 years.

185. "Questions and Answers About Seismic Gaps." Eos, v. 60, August 21, 1979, pp. 627-628.

A series of basic questions about the use of seismic gaps in earthquake prediction.

186. "Seismic Gaps: Some Questions and Answers." Earthquake Information Bulletin, v. 11, November/December 1979, pp. 221-223.

A description of the basic theory of seismic gaps in earthquake prediction.

187. Simon, C. "Aleutian Arc primed for a Great Quake." Science News, v. 123, June 18, 1983, p. 389.

Scientists have predicted a magnitude 8.6 earthquake will occur during the next twenty years in the Shumagin Gap in Alaska. A recent slip between the two plates should have caused a major earthquake but did not, increasing the chances for a future quake.

188. "Tectonics-Torn Italy." Science News, v. 118, December 13, 1980, p. 376.

Soviet and Italian scientists have mapped out the most dangerous spots in Italy for earthquakes using the seismic gap method. One of these danger zones experienced a magnitude 6.8 earthquake in November 1980.

189. West, Susan. "Alaskan Quake Potential." Science News, v. 115, June 9, 1979, p. 377.

A seismic gap exists in the Gulf of Alaska and a magnitude eight earthquake is expected within the next twenty years. The pattern of activity suggests Cape Yakataga as the most likely location.

Animals as Earthquake Predictors

190. "Animals Sense Change Before Quake Strikes." <u>Science Digest</u>, v. 83, March 1978, pp. 18-19.

Historical reports of animal behavior predicting earthquakes are now being examined scientifically. Researchers are trying to determine if animals can be used as short-term precursors and what mechanisms they use to detect earthquakes.

191. "Anxious Animals as Quake Precursors." <u>Science News</u>, v. 113, April 29, 1978, p. 278.

By interviewing residents of Willis, California following a recent earthquake, researchers have determined that animals can be used to predict earthquakes. One-third of those people interviewed reported unusual animal behavior prior to the quake.

192. Asher, Linda. "Growth of Seismic Premonitionology." <u>Psychology Today</u>, v. 13, February 1980, pp. 18-23.

Scientists are beginning to explore the use of animal behavior to predict earthquakes. A network of volunteers has been established in California to monitor animal behavior and to report it to a central location for analysis.

193. Buskirk, Ruth E., Cliff Frohlich, and Gary V. Latham. "Unusual Animal Behavior Before Earthquakes: A Review of Possible Sensory Mechanisms." <u>Reviews of Geophysics and Space Physics</u>, v. 19, May 1981, pp. 247-270.

Some animals are much more capable than humans of detecting certain kinds of geophysical stimuli, including seismic or acoustic waves, electric field changes, and gaseous emissions. Animals may well react to these changes before an earthquake, but there are not yet enough data to use animal behavior as a reliable earthquake precursor. Some recommendations are given for future research in geophysics and biology.

194. "Doomsday Headaches." <u>Science Digest</u>, v. 90, February 1982, p. 99.

A woman in Oregon has severe headaches that are well correlated with geologic events. It is theorized that she is acutely sensitive to electromagnetic waves produced by changes in the pressure of the earth's crust.

195. Kerr, Richard A. "Quake Prediction by Animals Gaining Respect." Science, v. 208, May 16, 1980, pp. 695-696.

It is still unclear if animals can detect an upcoming earthquake. A network of 1,200 volunteers has been put together to report unusual behavior. Some possible mechanisms for animal predictions involve detection of changes in low-frequency sound waves, electrical phenomena, or odors associated with gas emissions.

196. Ling-Huang, Shen. "Can Animals Help to Predict Earthquakes?" Earthquake Information Bulletin, v. 10, November/December 1978, pp. 231-233.

Many people reported unusual animal behavior prior to the Tangshan, China earthquake. This was similar to reports of other Chinese quakes. By observing and recognizing unusual behavior, future earthquakes may be forecast.

197. Lott, Dale F., Benjamin L. Hart, and Mary W. Howell. "Retrospective Studies of Unusual Animal Behavior as an Earthquake Predictor." Geophysical Research Letters, v. 8, December 1981, pp. 1203-1206.

Interviews were conducted following seven earthquakes in North, Central, and South America to determine if animals behaved differently before the quake. For comparison, interviews were also conducted in areas unaffected by the earthquakes. From this study, unusual behavior did precede one earthquake and did not precede four others.

198. Monagan, David. "How Animals Predict Earthquakes." Science Digest, v. 89, June 1981, pp. 92-95+.

One of the precursors used to predict the Haicheng, China earthquake was the unusual behavior of local animals. Some of the causes of this behavior may be changes in the earth's magnetic field, sound waves produced by microfracturing of rocks or the venting of gases, the scent of noxious gases released through the ground, or a change in the ion level of the atmosphere.

199. Rikitake, Tsuneji. "Biosystem Behavior as an Earthquake Precursor." Tectonophysics, v. 51, November 20, 1978, pp. 1-20.

Data on animal behavior indicate that they fall into two categories: moderately short precursors on the order of a few days and extremely short precursors on the order of a few hours. Few other precursors coincide with moderate animal behavior, but the extremely short range correlates with certain geophysical events. Animal behavior cannot be ruled out as an indicator of earthquakes.

200. "Sensing Quakes." Time, v. 113, January 29, 1979, p. 73.

Reports have been made throughout history of animals behaving oddly before an earthquake. A German scientist now believes that the pressure of the rocks releases charged ions into the air, causing unusual behavior.

201. Shaw, Evelyn. "Can Animals Anticipate Earthquakes?" Natural History, v. 86, November 1977, pp. 14-20.

Reports of unusual animal behavior are frequently heard following an earthquake. Chinese scientists used this as part of the evidence to predict the Haicheng earthquake. Researchers are studying how animals could detect a forthcoming earthquake and whether animals can be useful predictors.

202. "Sick? Might Mean a Quake is Due." International Wildlife, v. 13, May/June 1983, p. 24D.

Just as animals may sense an oncoming earthquake, humans may do so as well. A network of 25 volunteers in the San Francisco Bay area who appear to be sensitive to earthquakes have had an 84% success rate in predicting local earthquakes.

203. Tributsch, Helmut. "Do Aerosol Anomalies Precede Earthquakes?" Nature, v. 276, December 7, 1978, pp. 606-608.

Charged aerosol particles produced by the release of an underground electrostatic charge may be responsible for unusual animal behavior prior to earthquakes.

204. Tributsch, Helmut. "Nature's Earthquake Warnings." Atlas World Press Review, v. 26, March 1979, p. 54.

Animal precursors to earthquakes may be due to electrostatically charged aerosol particles in the atmosphere. The increased ion concentrations confuse the animals and cause them to behave strangely.

205. Tributsch, Helmut. "Seismic Sense: Something in the Air Panics Animals Before an Earthquake." The Sciences, v. 22, December 1982, pp. 24-28.

An earthquake in Friuli, Italy invoked research on the causes of unusual animal behavior preceding earthquakes. Using historical and anecdotal evidence, it is theorized that ions resulting from the release of static electricity within the earth may be responsible.

206. Tributsch, Helmut. When the Snakes Awake: Animals and Earthquake Prediction. Cambridge, Mass.: MIT Press, 1982. 248p. ISBN 0-262-20044-9.

Based on historical and anecdotal accounts, this book describes how animals have been used to predict earthquakes. The sensitivity of animals to the release of charged aerosol particles by rock under pressure is theorized to account for this behavior.

207. West, Susan. "Pet Seismologists." Science News, v. 117, June 14, 1980, p. 376.

In an effort to determine if animal behavior is a valid precursor to earthquakes, a team of 1,250 volunteers are watching animal behavior in the earthquake-prone areas of California. So far the results have been inconclusive.

The Social and Ethical Considerations of Earthquake Prediction

208. Aaronson, Steve. "Prediction of Tremblors Can Create New Hazards." Science Digest, v. 83, April 1978, pp. 64-70.

Issuing a scientifically credible earthquake warning could destroy the economy of the area. The longer the lead time, the deeper the resulting depression. This could lead to speculation and fraud by those trying to take advantage of the situation. Geologists need to determine if the harm of the prediction will be worse than the damage from the earthquake.

209. Blundell, D. J. "Earthquake Hazard Risk Mitigation." Contemporary Physics, v. 22, 1981, pp. 335-347.

Earthquake risk is a combination of three factors: earthquake hazard, loss potential, and vulnerability. In areas of high population density where the loss potential is high, it is important to reduce the vulnerability as much as possible by designing and building earthquake resistant buildings. At present we cannot accurately predict when an earthquake will strike.

210. Dickson, David. "Budget Cuts Cause Upheavals in Earthquake Research." Nature, v. 277, February 8, 1979, p. 421.

Congress has reduced the funding for earthquake hazard research from $26.4 million to $17.4 million. It may be a long time before money will be available for research on the social aspects of earthquakes.

211. "Earthquake Forecasts: The Perils of Prediction." Futurist, v. 13, June 1979, p. 233.

A forecast of an earthquake may do more damage to a community than the quake itself. A prediction for a given city may cause many residents to leave, inducing economic hardship. Inaccuracy in predictions will also cause the public to lose faith in warnings, resulting in a tendency to ignore the danger.

212. Fallows, James. "Living on the Fault Line." Atlantic, v. 248, October 1981, pp. 29-37.

There is a 50% chance of a great earthquake in southern California over the next forty years. This area is rapidly being developed and the potential for damage is tremendous. The Federal Emergency Management Agency has found that the nation is unprepared for a catastrophic earthquake and that we must begin to prepare for the inevitable results.

213. "Guidelines for Earthquake Predictors." Bulletin of the Seismological Society of America, v. 73, December 1983, pp. 1955-1956. Reprinted in: Earthquake Information Bulletin, v. 16, July/August 1984, pp. 167-169.

A set of recommended prefessional guidelines for scientists making earthquake predictions.

214. Hill, Mary. "California Seismic Safety Commission." Earthquake Information Bulletin, v. 10, November/December 1978, pp. 204-207.

The Commission was set up by the California Legislature to formulate earthquake policy and to reduce earthquake hazards. It works to make public buildings earthquake resistant and monitor the impact of earthquake warnings on state and local governments.

215. Hill, Mason L. and Lisa A. Rossbacher. "Seismic Risk: Case Studies--Real and Imaginary." Earth Science, v. 35, Fall 1982, pp. 19-24.

A study of the seismic threat from the Whittier, California fault. Geologic and seismic information may be used by both proponents and opponents of construction projects to support their positions.

216. Jackson, E. L. "Public Response to Earthquake Hazard." California Geology, v. 30, December 1977, pp. 278-280.

A recent survey shows that some government strategies in dealing with earthquake hazards discourage individuals and the public from taking their own actions. Ninety percent feel that preparation for future quakes is a government problem and only one percent have intentionally chosen a site away from hazardous ground.

217. Kiecolt, K. Jill and Joanne M. Nigg. "Mobility and Perceptions of a Hazardous Environment." Environment and Behavior, v. 14, March 1982, pp. 131-154.

An analysis of when and why people will move due to an earthquake threat. Surprisingly, very few people move away, despite warnings of an impending earthquake. Only those with a fear of earthquakes have a tendency to leave.

218. Kitazawa, K. "Earthquake Prediction and Public Response." Impact of Science on Society, v. 32, January/March 1982, pp. 29-37.

Greater attention to geophysical, geochemical, and biological anomalies is lending increased reliability to the methods and techniques of earthquake prediction. This technical advance must be accompanied by improved understanding of public behavior in the face of a warning if lives are to be saved when an earthquake occurs.

219. Lang, Berel. "Earthquake Prediction: Testing the Ground." Environmental Ethics, v. 5, Spring 1983, pp. 3-19.

The social consequences of earthquake predictions have implicated such varied environmental issues as land use, building codes, and economics. Lay members of the public are more directly involved in earthquake prediction than other scientific projects. Attention should be paid to the specific public and social values involved in earthquake prediction.

220. National Research Council. Committee on Seismology. Panel on Earthquake Prediction. Predicting Earthquakes: A Scientific and Technical Evaluation with Implications for Society. Washington, D.C.: National Academy of Sciences, 1976. 62p. ISBN 0-309-02527-3.

A summary of the state of the art of earthquake prediction and its effects on society. The time of successful earthquake predictions is just around the corner and can be achieved with an increase of government support for such research. The technology has reached the stage where it is important to study the sociological impacts of earthquake predictions.

221. National Research Council. Committee on Socioeconomic Effects of Earthquake Predictions. _A Program of Studies on the Socioeconomic Effects of Earthquake Predictions_. Washington, D.C.: National Academy of Sciences, 1978. 162p. ISBN 0-309-02789-6.

A comprehensive analysis of the effects of earthquake warnings on the economy and social structure of the affected region. Problems and reactions of the public, government, and the business community are all included, as well as the legal liability for earthquake predictions. Some suggestions for future research are made.

222. National Research Council. Panel on the Public Policy Implications of Earthquake Prediction. _Earthquake Prediction and Public Policy_. Washington, D.C.: National Academy of Sciences, 1975. 142p. ISBN 0-309-02404-8.

Earthquake prediction is still in the research stage but may soon be a practical weapon for fighting the damage caused by earthquakes. A list of recommendations for research is provided for such areas as disaster preparedness, prediction and warning, hazard reduction, emergency services, and dealing with the negative effects of earthquake predictions.

223. National Science Foundation. _Earthquake Prediction and Hazard Mitigation: Options for USGS and NSF Programs_. Washington, D.C.: Government Printing Office, 1976. 76p. Superintendent of Documents Number NS1.2:Ea7/2.

An explanation of the benefit to the public of increased research in the field of earthquake prediction.

224. Nigg, Joanne M. "Communication Under Conditions of Uncertainty: Understanding Earthquake Forecasting." Journal of Communication, v. 32, Winter 1982, pp. 27-36.

Information seeking relating to earthquake prediction takes four stages. In stage 1, individuals seek information for entertainment or general information. In stage 2, media attention increases and individuals discuss the prediction informally more frequently. In stage 3, the tone of the warnings changes and individuals seek more detailed information. In stage 4, individuals attempt to determine the importance of the warning on an individual basis and seek out expert advice on how to proceed.

225. Peppin, Richard J. "Earthquake Research." Mechanical Engineering, v. 103, February 1981, p. 71.

Earthquake prediction may have more negative effects than positive ones. When an earthquake is predicted, the economy of the area will suffer whether the quake occurs or not. Research on earthquake prediction should be halted.

226. "Predicting the Possibility of Natural Disasters Is a Terrible Problem." Center Magazine, v. 16, September/October 1983, pp. 10-11.

Although scientists have a moral responsibility to warn areas of impending earthquakes, the warning itself may have a negative effect on the economy.

227. "Public Policy and Earthquake Warning." Chemical and Engineering News, v. 53, September 1, 1975, p. 6.

The government will soon set up a panel of qualified scientists to review all earthquake predictions and to issue official earthquake warnings.

228. Raleigh, C. Barry. "Scientists' Responsibility for Public Information." Science, v. 213, July 31, 1981, p. 499.

Earthquake warnings may have a dramatic effect on the public. If a quake occurs after a vague warning, or if nothing happens following a specific warning, the predicting authorities may be held liable for costs or damages. Scientists need to be protected from legal action relating to earthquake predictions.

229. Rikitake, Tsuneji. "The Large-Scale Earthquake Countermeasures Act and the Earthquake Prediction Council in Japan." Eos, v. 60, August 7, 1979, pp. 553-555.

The Countermeasures Act was passed in 1978 to legislate earthquake warning procedures. Government officials are now empowered to make warnings and an extensive monitoring network has been established in high-risk quake areas.

230. Sorensen, John H. "Earthquake Prediction: A Socioeconomic View." Issues in Engineering, v. 3, July 1980, pp. 257-263.

The major socioeconomic factors involved in earthquake prediction are: 1. Uncertainty--no quakes can be predicted exactly. 2. Standards--efforts should be made to make sure that all predictions meet authorized criteria. 3. Frauds--Some individuals may make fraudulent predictions for personal gain. 4. Liability--Are scientists legally liable for damage claims? 5. Cost--Most individuals cannot afford to improve their property to meet earthquake-proof standards. 6. Priorities--When money and time is short, which buildings should be improved first? 7. Government--Should governments regulate predictions and to what degree should they carry out damage assessments or provide information?

231. Spall, Henry. "Seismology in South America: An Interview with Alberto Giesecke." Earthquake Information Bulletin, v. 12, March/April 1980, pp. 64-70.

Earthquake hazards are very real in Peru and often towns are rebuilt in the same location without considering these hazards. A seismic monitoring effort is currently underway to determine such high risk areas. We need to translate awareness of earthquake hazards into governmental actions.

232. Spall, Henry. "Sociological Aspects of Earthquake Prediction." Earthquake Information Bulletin, v. 11, May/June 1979, pp. 102-105.

Earthquake predictions affect people in many ways. Three strategies are generally followed: relocation from the affected area, a reallocation of resources within the area, or a reduction of resources being input into the area. Any earthquake warning will have a negative effect on the region and officials must work to contribute accurate information to minimize the loss.

233. "Understanding Earthquake Hazards." Earthquake Information Bulletin, v. 16, January/February 1984, p. 22.

The Federal Emergency Management Agency has developed three standards to be used in its earthquake education program. The information must be presented in a language familiar to the target audience, the program must be a dynamic process with continuous feedback from the recipients, and the program must include methods for evaluating its effectiveness.

234. U.S. Geological Survey. "Earthquake Prediction: Opportunity to Avert Disaster." Geological Survey Circular, no. 729, 1976, pp. 1-35.

The proceedings of a conference on earthquake warning and response. Topics covered include the science of earthquake prediction, methods of loss-reduction, and the social, economic, and political effects of earthquake predictions.

235. "USGS Revises Hazards Criteria." Eos, v. 65, March 27, 1984, p. 116.

The Geological Survey has revised its methods for issuing formal statements about earthquakes. The new system has two categories: a hazard warning (which is a formal notice) and an informal system for notifying officials about less critical information.

236. Wallace, Robert E., James F. Davis, and Karen C. McNally. "Terms for Expressing Earthquake Potential, Prediction, and Probability." <u>Bulletin of the Seismological Society of America</u>, v. 74, October 1984, pp. 1819-1825.

A list of suggested terms to be used when issuing hazard warnings. The two primary categories suggested are "long-term earthquake potential" and "earthquake prediction." Prediction is further subdivided into "long-term prediction" (in years), "intermediate-term prediction" (from weeks to a few years), and "short-term prediction" (within a few weeks). The use of standard terminology will assist geologists, government officials, and the public in assessing earthquake hazards.

237. Ward, Peter L. "Earthquake Prediction." <u>Mechanical Engineering</u>, v. 101, February 1979, pp. 44-46.

The age of prediction will be here in a decade or two. Accurate predictive methods are just around the corner. Whenever a scientist discovers information that may save lives, it should be released to the public even if it might be a false alarm.

Brian Brady and the Great Peruvian Quake Predictions

238. "Council Turns Down Peru Quake Forecast." Science News, v. 119, February 14, 1981, pp. 100-101.

The National Earthquake Prediction Evaluation Council has rejected a prediction of a major Peruvian earthquake. The council felt that the prediction was based on inconsistencies in the data and included unrealistic detail of the predicted time, place, and magnitude.

239. Joyce, Christopher. "Brady's Earthquake Doesn't Shake the Experts." New Scientist, v. 89, February 12, 1981, p. 393.

Brian Brady, a specialist in rock mechanics, has predicted a series of three strong earthquakes for Peru. The Geological Survey has called the work "vague and speculative." However, an earthquake that struck last December was predicted by Brady as a precursor to the main shocks.

240. Kerr, Richard A. "Earthquake Prediction Retracted." Science, v. 213, July 31, 1981, p. 527.

A prediction of three major earthquakes in Peru has been withdrawn. Because a preliminary earthquake did not appear, it is unlikely that the others will occur. Although the Geological Survey did not support this prediction, it was taken very seriously in Peru.

241. Kerr, Richard A. "Prediction of Huge Peruvian Quakes Quashed." Science, v. 211, February 20, 1981, pp. 808-809.

The National Earthquake Prediction Evaluation Council has rejected a prediction by Brian Brady of three massive Peruvian earthquakes. The council found that the methods used to determine the predictions were "vague and speculative." Despite pressure from the committee and the lack of expected foreshocks, the prediction has not been withdrawn.

242. "Official Earthquake Council Statement." Eos, v. 62, March 31, 1981, p. 129.

The text of the official statement by the National Earthquake Prediction Evaluation Council denouncing Brian Brady's Peruvian predictions.

243. "Peru Earthquake Prediction Update." Eos, v. 62, August 11, 1981, p. 618.

Because an expected foreshock did not occur, Brian Brady has altered his prediction of three major earthquakes in Peru.

244. "Peruvian Quake Prediction Modified." Science News, v. 120, August 1, 1981, pp. 70-71.

Brian Brady, who predicted three major earthquakes in Peru using his "inclusion collapse theory," has recently modified that prediction when the first quake did not occur.

245. "Prediction of Quakes Rejected." Eos, v. 62, March 31, 1981, p. 129.

The National Prediction Evaluation Council has not endorsed a prediction of three major earthquakes for Peru. The council could find no evidence to support the theory used to make the prediction.

246. "Quake Prediction Causes Own Shakes." Science News, v. 120, July 4, 1981, p. 5.

The June 28, 1981 earthquake predicted for Lima, Peru did not occur. Nevertheless, the prediction did create fear in the people of the city and much damage to the economy. Only time will tell if the theory behind this prediction is accurate or not.

Author Index

Name: Abstract Number(s).